IMAGES
of Aviation

NAVAL AIR STATION WILDWOOD

BOMBING SQUADRON
EIGHTY - FIVE

LT. W.W. PERRY, JR.

LT. T.R GAVIN

LT. E.F. GIBSON

LT. M.J. HEMBY

LT. W. GAUVEY, JR.

LT. T.C. THOMAS

LT. CDR. A.L. MALTBY, JR.
COMMANDING OFFICER

LT. A.G. SYMONDS, JR.
EXECUTIVE OFFICER

LT. (JG) R.W. ELMORE

LT. (JG) G.C. SHOEMAKER

LT. (JG) H.E. EAGLESTON, JR.

LT. (JG) J.F. HUNGERFORD

ENS. R.W. JONES

LT. (JG) J.F. LACINA

ENS. H.W. FORSGREN

ENS. J.A. LOCKE

ENS. R.H. WENT

ENS. R.W. MANN

LT. O.S. HARGETT, JR. LT. K.F. CALLAHAN ENS. C.M. CRUMP

ENS. M.J. MITCHELL

ENS. W.J. DOERING

ENS. C. TRUMP

ENS. G.V. EVEN

ENS. S.G. PAYNE

ENS. M.L. SKINNER

ENS. J.J. SCHERTING

The World War II pilots and officers of bombing squadron VB-85 pose in this composite with their SBD Dauntless fighter airplanes. The squadron trained at Naval Air Station Wildwood during the summer of 1944 before being sent to the carrier *Shangri-La* to fight in the Okinawa campaign during the spring of 1945. The squad was in combat on V-J Day, August 15, 1945. (Courtesy Naval Air Station Wildwood Aviation Museum archives.)

ON THE COVER: Its training completed, bomber squadron VB-86 begins leaving Naval Air Station Wildwood in September 1944, headed to hostilities in the Pacific theater. Comdr. Paul Norby returned from the Pacific to teach the new squadron dive-bombing tactics. The folded wings of the SB2C Helldiver bombers seen here, parked on the apron north of Hangar No. 1, allowed for compact storage on the flight decks of the navy's aircraft carriers. (Courtesy Naval Air Station Wildwood Aviation Museum archives.)

IMAGES
of Aviation

NAVAL AIR STATION
WILDWOOD

Joseph E. Salvatore, M.D. and Joan Berkey

ARCADIA
PUBLISHING

Published by Arcadia Publishing
Charleston SC, Chicago IL, Portsmouth NH, San Francisco CA

Printed in the United States of America

Library of Congress Control Number: 2009928149

For all general information contact Arcadia Publishing at:
Telephone 843-853-2070
Fax 843-853-0044
E-mail sales@arcadiapublishing.com
For customer service and orders:
Toll-Free 1-888-313-2665

Visit us on the Internet at www.arcadiapublishing.com

In memory of the 42 known navy airmen who lost their lives while training at Naval Air Station Wildwood during World War II.

CONTENTS

ACKNOWLEDGMENTS

Since the Naval Air Station Wildwood Foundation and Aviation Museum were founded in 1996, hundreds of volunteers have contributed tens of thousands of hours to preserve the history of the station's significant role in training fighter pilots and crewmen during World War II. From painstakingly rebuilding historic aircraft to restoring the historic Hangar No. 1 that houses the aviation museum, these volunteers have been the backbone of our efforts to tell the station's story. After the museum opened, many former World War II airmen and women who were stationed at Naval Air Station Wildwood have visited and generously shared stories, photographs, and other mementos of their service here. Unless otherwise noted, all images appearing in this book are from the museum's extensive archives, and many of them were donated to the museum by these airmen or their families.

Special thanks go to our editor at Arcadia Publishing, Erin Rocha, for her guidance and infinite patience as we encountered unexpected challenges while preparing this book. Nancy Zerbe and Stacy Spies of Arch II deserve special recognition for the nomination to the New Jersey and National Registers of Historic Places they prepared for Hangar No. 1; much of the information cited here comes from their work. Thanks, also, to John Fitzgerald, aviation enthusiast, for reviewing the draft manuscript and to Dr. Ronald Gelzunas who donated the Al Timer photograph collection. Lastly, we would like to thank the Wildwood Historical Society and the Cape May County Historical and Genealogical Society for sharing their files on the air station during our research.

INTRODUCTION

The Civil Aeronautics Act of 1938 set in motion a nationwide movement to develop airports for national defense purposes, wherein sites for military airfields were to be selected with regard to future civilian aviation potential, cost, and efficiency. By building on land offered by local governments or augmenting extant civil airports, the Civil Aeronautics Administration (CAA) also aided in the wartime economy of communities positioned near the bases. Of the 86 facilities built in the eastern states, 43 were constructed at existing municipal or civilian airfields.

Such was the case with the land appropriated for U.S. Naval Air Station Wildwood (USNASW), which had been purchased by Cape May County with a $25,000 bond issue raised for the establishment of a county airport 4 miles outside Wildwood. Three 1,000-foot runways were then constructed in Wildwood, for use by the army or navy, by the CAA upon the outbreak of World War II in 1939. The navy then leased the airport for $1 per acre and acquired an additional 38 acres on which to build a base. USNASW consisted of 942.751 acres of land situated in the center of the lower third of the Cape May Peninsula, in Lower Township.

With the growing threat of the United States' involvement in the war, Congress granted an 80 percent increase in funding for construction of naval air combat structures in June 1940. By April 1942, five months after the attack on Pearl Harbor, agreements were finalized to expand the Cape May County Airport and on April 1, 1943, it was commissioned as the United States Naval Air Station Rio Grande. Due to difficulty in obtaining adequate mail, telegram, and telephone service resulting from confusion with Rio Grande, Texas, the station was redesignated as USNASW two months later.

At the time of commissioning, the station was directly under the authority of the commandant, Fourth Naval District, and was assigned the mission of providing facilities and services for the operation of fleet units under commander, Fleet Air, and Quonset Point. Construction of Hangar No. 1 began in October 1942 as part of a project to support the initial group of 108 officers, 1,200 men, and 72 planes.

In August 1943, the CAA approved construction of three 2,500-foot runways in a vacant field 18 miles outside USNASW in Woodbine, New Jersey, to compensate for an increase in field traffic at both USNASW and Naval Air Station (NAS) Atlantic City. This outlying field (OLF) was utilized by planes from both stations and administered by USNASW until December 1943, when it was transferred to NAS Atlantic City located some 35 miles to the northeast. At that time, USNASW was granted permanent assignment of three existing 5,000-foot runways in an OLF 35 (air) miles away in Georgetown, Delaware. Improvements made by USNASW at the Georgetown OLF included additional increased fuel storage, catapult and arresting gear, a building, and runway lighting. The Georgetown OLF was used throughout the active history of USNASW exclusively for day and night carrier landing practice.

USNASW went through several phases of wartime operation. Upon its commissioning on April 1, 1943, USNASW was intended to serve as a training ground for the formation of complete air groups operating from aircraft carriers. By June 1943, USNASW was being used only for the training of dive-bombing squadrons, with the other common air carrier group constituents (fighting, bombing, and torpedo training) relegated to other stations.

Consideration for an appropriate dive-bombing training base includes level, well-drained ground available from any approach, and suitability for takeoffs and landings under instrument. Sited on a peninsula surrounded on three sides by water—the Atlantic Ocean and the Delaware Bay—USNASW was ideally sited for this specialized air warfare training. The terrain was flat, the weather offered almost-ideal flying conditions year-round, and practice missions could be carried out over both land and water.

With an expansion in late 1943, the number of aircraft assigned to the station rose to over 200. Except for the inclusion of a single fighter squadron from September to November 1944, USNASW trained only dive-bombing squadrons until January 1945. At that time, training of dive-bombing squadrons relocated to NAS Manteo, with only fighter-bomber (VBF) squadrons based at USNASW.

The training of fighter-bomber squadrons was improved at USNASW with the installation of two rocket targets on the southwestern shore of the Delaware Bay. Within four months, the dive-bombing squadrons returned to USNASW as the facilities at NAS Manteo proved insufficient for training. From June to August 1945, "a perfected and concentrated program of training had been developed and was in full operation" for both dive-bomber and fighter-bomber squadrons at USNASW. In July 1945, facilities at the station were further improved when a 20-millimeter school and an aircraft carrier navigation program were developed.

With key advantages of increased aim accuracy and a safer bomb release, dive bombing had been integrated into U.S. Naval operations as early as 1928. By the 1930s, a typical air carrier group consisted of bombing (VB), fighting (VF), scouting (VS), and torpedo (VT) squadrons. This arrangement continued into World War II, although the units did evolve as battle and circumstance necessitated. While USNASW originally trained complete air groups, after two months it was designated for only dive-bombing squadrons—possibly preferred for its coastal, inlet setting.

The training of dive-bomber pilots was particularly dangerous, and in its two years of operation USNASW suffered 194 accidents, including 42 known deaths. Prior to arrival at USNASW, pilots spent three months in basic training and several more months in ground school, squadron, and preflight training. Training received at USNASW included night and instrument flying, takeoff and landing from the deck of a carrier, and diving at a 70-degree angle within 200 to 300 yards of a target with bomb release and pull out. The seven elements of the dive-bomb attack included in training at USNASW were approach, break, dive, drop (shot), pullout, withdraw, and rendezvous.

Dive-bomber aircraft differed from other military aircraft in two distinct ways: dive brakes were used to slow diving speed to allow the pilot more time to accurately position the bomb, and the aircraft had a hinged bomb rack under the fuselage so the released bomb would clear the arc of the propeller. Initial dive-bombing planes utilized at USNASW were the Douglass SBDs, also known as "Dauntless." Known to be durable and reliable, the SBD was tested for use by the navy in early 1940. By the spring of 1942, Douglass increased its production of the SBD for the navy after the plane proved to be successful in the new carrier-versus-carrier battles of the war.

Though integral to the carrier-based bombing force for the first two years of the war, by late 1942 the SBD was already being phased out in favor of the Curtiss SB2C, also known as "Helldiver." It would be the last line of aircraft developed by the U.S. Navy specifically for the role of dive bombing, and it went through many difficulties in development and production. Its comparative success versus the SBD remains contested, although the SB2C was responsible for the destruction of more Japanese targets than any other dive bomber. SB2Cs were used in combat for the first time in the Second Rabaul Strike on November 11, 1943.

In January 1945, fighter-bombing (VFB) squadrons began to be established and trained at USNASW using the Vought F4U Corsair. While joint bomber/fighter missions were particularly vulnerable to Japanese fighters due to typically deficient training in glide-bombing tactics by fighter pilots, the Corsair offered many advantages in battle. It could carry the same load as the SB2C, but over a shorter range and faster. Performance was typically very high, and the plane could be used with or without bombs and rockets.

Unique features of this low-wing monoplane include a propeller of unusually large diameter and inverted gull wings, which kept the main landing gear short and retractable straight back, with a fuselage high enough to compensate for the large blades. The Corsair was first used in combat in Guadalcanal on February 11, 1943, and became integral to success in the Pacific arena by 1944. It stayed in production longer than any other U.S. fighter of the period and was credited with an 11:1 ratio of kills/losses against the Japanese.

Activity at USNASW peaked in October of 1944, with 16,994 takeoffs and landings and accommodation of as many as 200 planes. The station's enlisted personnel complement upon its opening was 122 and had grown to 362 by the time the station closed. In April 1946, the U.S. Navy turned the station over to the War Assets Administration (WAA) for disposition. Attempts by various county officials to prevent the station's closure that year were unsuccessful and 79 of the 126 original buildings were sold and moved off the site.

The federal government turned the remainder of the buildings, including Hangar No. 1, over to the county for use as the county airport on December 1, 1947. Through the years, most of the naval air station buildings that remained were torn down and replaced by more modern structures for use by Lower Township and private businesses. Today only 12 of the original air station buildings remain, and as a group they have been determined by the New Jersey Historic Preservation Office to comprise a historic district that is eligible for listing in the New Jersey and National Registers of Historic Places.

Hangar No. 1, the largest of the original structures, had various tenants, some briefly operating profitable independent airlines after the war. However, the oversized structure lay vacant and abandoned from 1992 until 1997 when the Naval Air Station Wildwood Foundation, under the leadership of retired orthopedic surgeon Dr. and Mrs. Joseph E. Salvatore of Cold Spring, rescued the historic building, procured grant money to restore it, and established an aviation museum and memorial to the navy airmen killed while training there during World War II.

The following units were located at USNASW during World War II, with basic information cited for each. Condensed genealogies for many of them can be found in *The History of the Fourth Naval District, Part VI*. Acronyms used include: Scouting Squadron (VS), Bombing Squadron (VB), Fighter Squadron (VF), Torpedo Squadron (VT), Composite Aircraft Squadron (VC), Carrier Air Group (CAG), Carrier Aircraft Service Unit (CASU), and Fighter-Bombing Squadron (VBF).

Carrier air groups at USNASW were as follows: CAG 16—reformed at Wildwood August 1944, consisted of VF-16 and VB-16, CAG-16 departed for Naval Auxiliary Air Station Oceana November 1944; CAG 30—established at USNASW 4/1/1943, consisted of VC-30 and VF-30, departed for Naval Air Station Norfolk July 1943; and CAG 81—reported aboard at USNASW August 1945.

Carrier aircraft service units at USNASW were as follows: CASU-23 DETACHMENT—assigned to support fleet air detachments April 1943, transferred to Naval Air Station Atlantic City August 1943; and CASU-24—arrives USNASW August 1943.

Squadrons at USNASW were as follows: VB-2—VB-14 arrives at USNASW June 1943, merges with VB-15 June 1943, designation changed to VB-27 July1943, in Naval Air Station Quonset Point July1943; VB-3—established at USNASW May 1945, departed for NAAS Oceana July 1945; VBF-3—established at USNASW May 1945, departed for NAAS Oceana July 1945; VB-4—reported aboard at USNASW July 1945, then assigned to Tarawa (CV-40) in the

Western Pacific in 1946; VBF-4—reported aboard at USNASW July 1945; VB-7—established at USNASW January 1944, departed for NAAS Oceana March 1944; VB-10—reformed at USNASW September 1944, departed for Naval Auxiliary Air Field Groton December 1944; VB-13—established at USNASW November 1943, departed for NAAS Oceana January 1944; VB-14—arrives at USNASW June 1943, merged with VB-15 June 1943, designation changed to VB-27 July 1943; VB-15—arrives at USNASW June 1943, merged with VB-14 June 1943; VB-20—reformed at USNASW April 1945, departs for Naval Air Station Edenton June 1945; VBF-20—established at USNASW April 1945, departs for NAS Edenton June 1945; VC-52—established at USNASW September 1943, transferred to NAS Quonset Point; VBF-75—VBF-75A reported aboard at USNASW June 1945, redesignated VBF-75 August 1945, departed for Naval Auxiliary Air Station Chincoteague August 1945; VBF-75A—reported aboard at USNASW June 1945, redesignated VBF-75 August 1945; VBF-75B—decommissioned August 1945; VB-80—established at USNASW February 1944, departed for NAAS Oceana March 1944; VB-81—established at USNASW March 1944, departed for Naval Auxiliary Air Field Otis May 1944; VBF-81—reported aboard at USNASW August 1945; VB-82—established at USNASW April 1944, departed for NAAS Oceana June 1944; VB-83—established at USNASW May 1944, departed for NAAF Otis July 1944; VB-85—established at USNASW May 1944, departed for NAAF Otis August 1944; VB-86—established at USNASW June 1944, departed for NAAF Otis October 1944; VB-87—established at USNASW July 1944, departed for NAAS Oceana September 1944; VB-88—established at USNASW August 1944, departed for NAAF Otis November 1944; VB-89—established at USNASW October 1944, departed for NAAS Oceana January 1945; VB-92—established at USNASW December 1944, departed for NAAF Groton February 1945; VB-93—established at USNASW December 1944, departed for NAAF Otis February 1945; VB-94—established at USNASW November 1944, departed for NAAS Oceana January 1945; VB-95—established at USNASW January 1945, departed for NAAS Oceana March 1945; VF-95—departed for NAAS Oceana January 1945; VBF-95—departed for NAAS Oceana March 1945; VB-97—established at USNASW November 1944; VB-150—established at USNASW January 1945, departed for Naval Auxiliary Air Station Edenton March 1945; VBF-150—reported at USNASW January 1945, departed for NAAS Edenton March 1945; VB-151—established at USNASW February 1945, departed for Naval Auxiliary Air Station Manteo February 1945; VBF-151—established at USNASW February 1945; VBF-152—established March 1945, departs for NAAF Groton May 1945; VB-153—departed for NAAS Oceana June 1945; VBF-153—established March 1945, departed for NAAS Oceana June 1945; and VB-306—reported at USNASW November 1943, departed for San Diego December 1943.

One

FROM FARMLAND TO STATION IN MONTHS

This aerial photograph, taken August 9, 1943, and looking east from an altitude of 6,000 feet, shows the almost-completed naval air station nine months after construction began. Seen in the center of the photograph is the largest of the station's buildings, Hangar No. 1, which is the size of two football fields. Located nearby are the operations building with control tower, the storehouse, the mess hall, several barracks, and the post office. (Courtesy of the National Archives.)

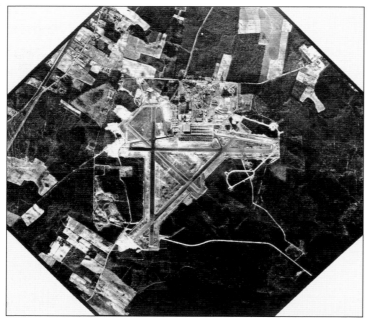

Taken on March 14, 1944, this octagonal-shaped aerial photograph shows the 942-acre facility with its buildings and four runways set amid a patchwork of farms and orchards. At the time, Cape May County's economy was largely dependent on agriculture, and the farms supplied produce to the county's many island resort communities. (Courtesy of the National Archives.)

When this photograph was taken in 1944, Hangar No. 2 was under construction and its framework is seen near Hangar No. 1. More barracks, including one for the Women Accepted for Volunteer Emergency Service (WAVE), have been erected to handle the increased number of airmen being trained here. Also built by this time are tennis, basketball, and volleyball courts, as well as an athletic field, all at the west end of the air station.

All of the station's 126 buildings were built by the time this 1945 photograph, which looks to the southeast, was taken. The immense Hangar No. 1 and half-Hangar No. 2, both surrounded by warplanes parked on the aprons, are seen in the center left. To the left of them is the operations building with its four-sided control tower that overlooks the four runways.

The official miniature plot plan for Naval Air Station Wildwood shows the 900-plus acre facility had a true north/south orientation with four runways placed to the north of the support buildings. The south border of the station is defined by Breakwater Road; the public entrance was also located on this road.

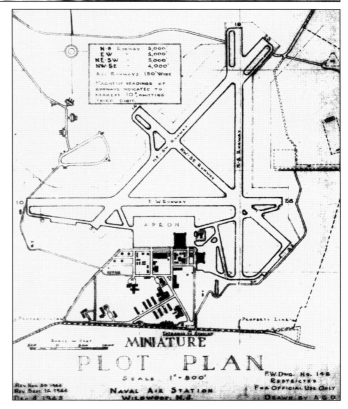

This stunning aerial photograph taken in 1945 looks northwest over the air station and towards the Delaware Bay. The bay, located just 2 miles to the west, played an important role in training the airmen for over-water diving and bombing tactics. In fact, most of the practice targets used by the fighter pilots were located in the water or along the marshes and beaches that dotted the Delaware Bay coastline in both New Jersey and Delaware. The network of dirt roads around the runways provided access to the ordnance storehouses that were kept at a distance from the station for safety reasons. The small town fronting the bay is known today as Villas and was developed in the 1920s, two decades before the station was built. (Courtesy of the National Archives.)

Medal ceremonies and personnel inspections were often conducted on the apron between Hangar No. 1 and the operations building. The operations building, seen here in the background, was one of the first buildings to be erected when construction began in 1942. The six nearly square buildings close to the terminal are not on the official site plan, but they may have served as temporary housing or as ready rooms.

This 1945 photograph shows Hangar No. 2 to the left, the operations building in the center, and Hangar No. 1 to the right. The westernmost two bays of the operations building, seen to the far right, contained a four-engine firehouse responsible for putting out fires on the runways and aprons. A fire truck is parked in front of one of the bays. (Courtesy of the National Archives.)

The interior of the control tower was filled with instruments considered modest by today's standards but was state-of-the-art then and included radio communications equipment, a dial telephone, and indicators of wind speed and direction. A pair of binoculars, sitting on the counter, allowed controllers to monitor the runways. Parked on the apron out front are Helldivers, Corsairs, and support aircraft. (Courtesy of the National Archives.)

Both the control tower and the one-story south wing of the operations building were topped with a variety of radio antennae used for communications between pilots and the command center. The small building in the foreground is the pump house. The operations building and the pump house (now a boiler room) still stand today. (Courtesy of the National Archives.)

This 1945 view shows the addition of an enclosed flight of stairs against the west wall to reach the control tower. Previously, a flat two-story ladder against that wall provided the only access to the tower and offered no protection in inclement weather. The four-sided tower was blown off in a storm in 1962 and was never replaced.

The five-story tower atop the operations building was lined with windows on all sides, allowing staffers to monitor the takeoffs and landings of aircraft, as well as the arrivals and departures of visitors, officers, and personnel. Seen behind the building is Hangar No. 1, which had its own, smaller control tower on the northeast corner.

Hangar No. 1, seen here looking east, was the largest building within the air station complex. The size of two football fields, it has 4,440 windowpanes and telescoping doors that slide open on tracks to allow ingress and egress of warplanes needing servicing or repairs. Lined up in front of the hangar are SNJs, training planes used at the station.

A TBM-3E Avenger fighter plane (left) and a multi-passenger Beechcraft are parked in front of the east side of Hangar No. 1. Seen to the far right is the hangar's small control tower. The hangar's telescoping doors, topped with multi-pane windows, were painted battleship gray while the rest of the building was painted white, colors that remain in use today.

Hurricanes were not common seasonal occurrences, but when the great Atlantic hurricane of 1944 threatened the eastern seaboard in September, as many of the station's planes as possible were tightly packed in Hangar No. 1 for protection. The hurricane did not make landfall in New Jersey but did bring high winds along the Jersey coastline. The hooked hoist, seen hanging here, was used to lift engines and propellers needing repair.

On November 1, 1943, Rear Adm. Calvin T. Durgin (right) and Lt. Comdr. Edwin G. Asman were photographed inside Hangar No. 1 in front of the athletic office. Because of a shortage of structural metal, which was directed for other war purposes, the hangar was built almost entirely of wood. Seen here is one of its large support posts, standing on a concrete pad and bolted to the floor. (Courtesy of the National Archives.)

A Grumman seaplane is parked in front of Hangar No. 2, which was built as a half-hangar in 1945. Parked on the apron to the side are Corsairs and Helldivers. The number of vacant spaces, marked by chocks, suggests that many of the station's planes were being used for training the day the photograph was taken.

Not surprisingly, the station's water tower was among the first structures to be erected. It originally stood near Hangar No. 1 and was torn down in the late 20th century. The white building in the distance is Hangar No. 1; the shorter, darker-colored building in front of the hangar is the armament repair shop.

Aviation Utility Shop A, seen here in this photograph taken May 1, 1944, featured a four-story vented "tower" that was used for drying out fliers' parachutes. The brick smokestack to the right marked the location of Heating Plant B. The white building in the background is the upper story of Hangar No. 1, which housed office space. (Courtesy of the National Archives.)

Taken August 24, 1945, just nine days after V-J Day was declared, this photograph shows navy captain John Donovan inspecting navy men on the apron between Hangar No. 1 and the operations building. Corsairs and SNJs, their wings folded for parking, stand in the background while several fire trucks appear to be on stand-by for any emergency that might occur.

This nondescript gatehouse welcomed visitors and personnel alike to the air station. Like many of the buildings erected there, it was covered with panels of transite, a hard fireproof composite material made from asbestos. Seen to the left behind the gatehouse are the one-story beauty parlor and the two-story ships service and post office building, none of which are standing today. (Courtesy of the National Archives.)

Two-story barracks for enlisted men lined both sides of Saratoga Road at the station. Four were built first, followed by another three as the number of airmen being trained increased dramatically in 1944. Like Hangar No. 1, the gatehouse, and many other buildings at the station, the barracks were covered with transite panels. To the left is Pump House No. 2.

The rotating beacon, which guided airmen to the station, was sited just south of both hangars, atop a seven-story wood-framed tower. It is not known what purpose the gangplank-type walkway served, but it may have been used for observing takeoffs and landings occurring on the N-S runway located about 330 yards to the east and seen in the background. (Courtesy of the National Archives.)

The target lookout tower stood along the shore of the Delaware Bay, in nearby Green Creek, located about 4.5 miles to the northwest of the station. Men positioned inside the tower were charged with observing bombing practice and reporting accuracy statistics and other information back to the pilots as they trained. (Courtesy of the National Archives.)

The transmitter shack, fronted by tall antennae, provided much-needed radio communication between the airmen and the base station. Its original location is not known, but it appears to have been built along the shores of the newly built (1942) Cape May Canal, which was located about 3 miles to the southwest. (Courtesy of the National Archives.)

This ammunition dump was one of about a dozen located on the west side of the air station used to store fuses, detonators, and small arms. Typical of munitions buildings at the time, it was mounded with dirt to protect the ammunition and contain any potential explosions. A ventilation pipe can be seen emerging from the roof at one end. (Courtesy of the National Archives.)

Located in the heart of the air station on the west side of Hangar No. 1, this ammunition locker may have been a temporary structure because it is not identified on the official site plan. Similar to the others, it was covered with earth for added protection should an explosion occur. Seen in the distance is the rotating beacon, the smokestack for the heating plant, and the firehouse/garage. (Courtesy of the National Archives.)

German Prisoners of War (POWs) performed a variety of duties at the air station and were housed in a former World War I army camp facility at Dias Creek, now the site of the county Mosquito Control Commission. During the Depression, the camp housed some 300 youths who helped clear drainage ditches to alleviate the mosquito population, a task later also assigned to the German POWs. (Courtesy of the National Archives.)

Two

THE AIRCRAFT
THEY FLEW

Two unidentified airmen, a pilot with a gunner in the rear, fly their Curtiss SB2C Helldiver over the farm fields of Cape May County in this 1945 photograph. Although not without its faults, the dive-bombing Helldiver was much faster than the SBD Dauntless it replaced and could keep up with the cruise speed of other fighter airplanes. It also featured folding wings, twin 20-millimeter cannons, and an internal bomb bay.

An SB2C Helldiver taxis down one of the runways at Naval Air Station Wildwood. Its folding wings allowed rows of planes to be tightly parked on the narrow deck of an aircraft carrier. The aircraft was manufactured by Curtiss, a company that eventually became a specialty supplier to the aircraft industry. The company produced 29,269 aircraft during the war, of which 7,140 were Helldivers.

A pilot, identified only as Hutchinson, poses in front of an F4U Corsair parked on the apron at Naval Air Station Wildwood. The airplane's speed, firepower, maneuverability, and ruggedness cause many to rate it with the Mustang as the best fighter plane of World War II. It featured an 18-cylinder double-radial engine capable of 2,250 horsepower with speeds nearing 400 miles per hour.

Support aircraft, like this twin-engine Beechcraft, were also commonly seen at Naval Air Station Wildwood. The Beechcraft was not a fighter plane; rather, it was used to transport executives and officers to various naval stations and airfields on an as-needed basis. Several models were produced during the war and some were used specifically for training military pilots, bombardiers, and navigators.

The Grumman F-6F Hellcat, like the one seen here on a runway at Naval Air Station Wildwood, was a carrier-based fighter aircraft developed to replace the earlier F4F Wildcat used by the navy. The Hellcat and Vought's F4U Corsair were the U.S. Navy's primary fighter aircraft during the second half of World War II.

This Northrup P-61 is photographed here at Naval Air Station Wildwood on August 8, 1945, shortly before the war ended. The P-61, known as the "Black Widow," was a twin-engine, twin-tail/rudder aircraft flown as a night fighter mostly by the U.S. Army Air Forces squadrons. The P-61 featured a crew of three: pilot, gunner, and radar operator. It was armed with four 20-millimeter forward-firing cannons and four 12.7-millimeter machine guns.

Navy men gather to watch this Curtiss SB2C Helldiver on one of the air station's runways. Two men are doing something on the pavement underneath it while three others stand or squat nearby, perhaps ready to assist. The Helldiver was a carrier-based dive-bomber aircraft that was one of the mainstays of the station's training fleet.

The F4F Wildcat, manufactured by Grumman, was used by both the U.S. Navy and the British Royal Navy when introduced in 1940. It had a top speed of 318 miles per hour that was bested by the more nimble Japanese Mitsubishi Zero, which could attain 331 miles per hour. The Wildcat was replaced by the Hellcat, which significantly outperformed the Zero.

Flying in formation during maneuvers at Naval Air Station Wildwood, these Curtiss SB2C Helldivers could carry 1,000 pounds of bombs mounted to the rocket hardpoints affixed to the bottom of the airplane's wings. It could also carry an internal torpedo and featured two fixed forward 20-millimeter cannons and machine guns in the rear cockpit.

A Curtiss SB2C Helldiver taxis near the refueling station in front of the air station terminal building in this undated photograph. The Helldiver had a reputation for being difficult to handle at low speeds but was responsible for destroying more Japanese targets than any other aircraft during World War II. It carried two men and had a top speed of 295 miles per hour.

A Mitsubishi Zero, the most famous Japanese fighter plane of World War II, was captured and brought to the air station for close scrutiny. The Zero proved a lethal plane against the Allied Forces in the first half of the war, having an exceptional range of 1,600 miles combined with unmatched maneuverability and firepower.

The captured Japanese Zero, seen here in front of Hangar No. 1 in 1944 or 1945, was test flown by Naval Air Station Wildwood airmen. The Zero was designed for attack but had no self-sealing tanks or armor plate, which contributed to high losses during combat. The Zero's early dominance was short-lived when the United States developed counter-attack tactics and new aircraft later in the war.

Some of the U.S. Navy's best-known fighter airplanes are shown here, lined up on the immense apron outside of Hangar No. 1 in 1944. They include Wildcats, F4U Corsairs, SNJs, SB2Cs, and some Howards, all of which were flown in a variety of naval combat and transport missions during the war. Those with folding wings were the easiest to store on the limited space of an aircraft carrier's flight deck. (Courtesy of the National Archives.)

Navy pilot William Lindgren poses next to an F4U Corsair with its bomb bay doors open. The Japanese nickname for the plane meant "whistling death," partly because the engine had a distinctive sound caused by the wing inlets for engine air. One of these is seen directly behind the pilot's head.

In March 1945, rows of Curtiss SB2C Helldivers line the apron in front of Hangar No. 1, while Corsairs taxi out to the runway. A line service truck with several bombs on its cargo deck is parked in front of No. 15086. Seen in the distant left is the fuse and detonator building, constructed far away from the main buildings for safety reasons. (Courtesy of the National Archives.)

A group of Douglass SBD Dauntless dive bombers fly over the Cape May coast in this undated photograph. The Dauntless was the U.S. Navy's main dive bomber from mid-1940 until it was augmented with the SB2C Helldiver in late 1943. Dauntlesses that arrived at Hawaii from the USS *Enterprise* were caught in the attack on Pearl Harbor.

An SB2C (above right) and a Corsair prepare to lay a smoke screen over Cape May County near the Delaware Bay in August 1945. Both airplanes were common sights in the air during training maneuvers. An antenna was strung from the "pole" sticking up from their fuselages and allowed them to communicate with each other and with the air station.

SB2C Helldivers practice aircraft carrier takeoff in May 1944. The pilots had to practice cueing and had to learn the appropriate time to unfold the airplane's wings. Often the outline of an aircraft carrier flight deck was painted on one of the runways, giving the fliers a clear vision of the limited space within which they had to maneuver. (Courtesy of the National Archives.)

The station insignia was designed by Graydon R. Locke of the station's Carrier Aircraft Service Unit (CASU) Operations unit. His winning entry in the insignia design contest, which depicts a Corsair taking off from an aircraft carrier, earned him a $10 award. It is not known how many designs were entered, but the decision by the judges was unanimous.

The afterwards folding wings seen on the Grumman Hellcats lined up in the forefront of this photograph were typical of that company's fighter planes during World War II. A gasoline truck and an oil truck are ready to service the many aircraft standing on the apron. Behind the Hellcats are several rows of Corsairs, along with a few SB2Cs and a Wildcat. (Courtesy of the National Archives.)

Most of the squadrons had their own logo. This one, for bombing squadron VB-80, featured an SB2C clutching a bomb with a devil for a pilot. VB-80 was established at Naval Air Station Wildwood on April 1, 1944, and was sent to the Pacific in 1945 where it was involved in operations over Tokyo, Iwo Jima, and Okinawa before the war ended.

Four SB2Cs, with engines revving, practice carrier takeoffs at the head of the flight line on March 6, 1944. The white outline drawn on the apron simulates the size and configuration of a carrier deck; it was used to orient the pilots to real-life flying conditions on board an aircraft carrier. Servicemen are stationed by each aircraft's landing wheels, ready to pull the chocks so the airplane can advance forward and take off. Placed at the end of each row of aircraft is a fire extinguisher, which would be used if an engine caught fire when starting, something that happened occasionally. The navy pilots had to learn the meaning of various signal paddles and flags used by the deck crew for takeoffs and landings. Many carrier landing and take-off accidents occurred when the pilot misinterpreted a signal from the deckhand. (Courtesy of the National Archives.)

The Grumman F4F Wildcat was the first fighter used by navy airmen in World War II. The wheels deployed from the fuselage rather than from the wings as in later fighter airplanes. Although the Japanese Zero handled better, the Wildcat had relatively heavy armor and self-sealing tanks that allowed it to survive more damage than its enemy counterpart. It also had a homing device that allowed pilots to find their carriers in poor visibility.

Two rows of SNJs are parked on the apron in front of Hellcats, Corsairs, and SB2Cs. SNJs were single-engine advanced trainer aircraft used to train army, navy, and Royal air forces during World War II. The United States continued to use the fighter trainer until the end of the 1950s.

Although never a part of the fleet at Naval Air Station Wildwood, this Consolidated PBY-6A visited the station sometime in 1945. A flying boat used by the navy, it could be equipped with depth charges, bombs, torpedoes, and machine guns and was one of the most widely used multi-role aircraft of World War II. Although slow and ungainly, it was highly reliable.

The twin-propeller PBY (center) draws a crowd of onlookers where it is parked in front of Hangar No. 2 in 1945. Nearby are two Howards, which were support airplanes used to transport officers and personnel. Also seen are a Beechcraft, another support airplane, and a Corsair. PBYs saved the lives of thousands of aircrew downed over water during the war.

Navy pilot Albert Timer poses on the wing of his SNJ while stationed at Naval Air Station Wildwood. Built by North American, these two-place planes were flown extensively by the navy. Though best known as a trainer, it was also used in combat in World War II and in the early days of the Korean War. It had several nicknames, among them Pilot Maker, Old Growler, J-Bird, and Mosquito.

Four Curtiss SB2C Helldivers fly in formation over the farms and fields of Cape May County in this undated photograph. They were piloted by Barnes (5), Hein (6), Hugel (8), and Wilson (9), identified only by their last names. The photograph may have been taken by the pilot or gunner in plane 7.

Two SNJ trainer aircraft seem to float in the clouds over Naval Air Station Wildwood. Made by North American (now Boeing), the SNJ two-place advanced trainer was the classroom for most of the Allied pilots who flew in World War II. It was designed as a transition trainer between basic trainers and first-line tactical aircraft and was used to train several hundred thousand pilots in 34 different countries over a 25-year period. Although not as fast as a fighter, it was easy to maintain and repair, had more maneuverability, and was easier to handle. It could roll, Immelmann, loop, spin, snap, and vertical roll. It was designed to give the best possible training in all types of tactics, from aerial dogfighting to ground strafing. It also contained almost every device that military pilots had to learn to operate, including blind-flying instrumentation, bomb racks, fixed and flexible guns, and cameras.

Three

TRAINING FOR WAR

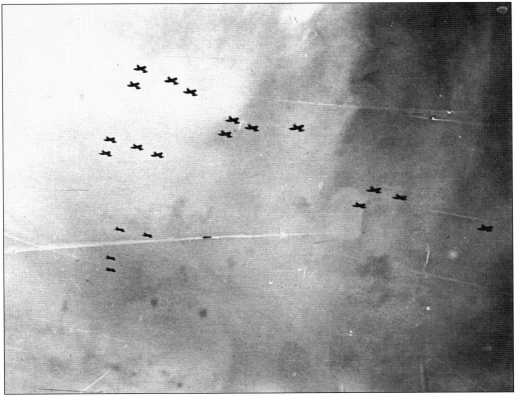

These Corsairs are flying as a squadron in formation, much like geese follow one leader, over the skies of Cape May County. The purpose of flying in formation during combat was to provide concentrated firepower against enemy fighters; also, the tighter the formation, the more difficult it was to be penetrated by enemy planes.

This fighter plane, probably an SNJ, is laying a smoke screen over the runways at Naval Air Station Wildwood in March 1945. Smokes screens importantly allowed a warplane to mask its movements, as well as the movement of other members of the fighter aircraft squadron, from airborne and seaborne enemies. (Courtesy of the National Archives.)

Five stories high, the skeet tower provided another important component of training at the air station. Clay discs were launched from the tower into the air at high speed from a variety of angles, teaching machine gunners the principle of leading and timing on a flying target. The tower was located on the northeast side of the station, well away from the runways, barracks, and support buildings. (Courtesy of the National Archives.)

Several fighter planes flying in formation could be effectively hidden in the smoke screen seen here laid by one plane flying over the coastline of the Delaware Bay. Often smoke screens were created by spraying oil over the hot exhaust of a plane's engine. In the Second Battle of the Atlantic in World War II, smoke screens were used by Allied destroyer escorts to mask the presence of merchant ships from German U-boats. Similarly, fighter aircraft were frequently used to lay smoke screens to hide the track and direction of warships during encounters at sea. It was common for one aircraft to lay a smoke screen for another type of aircraft depending on the end result needed. This photograph also shows the relative desolation of the coastline along the bayside of Cape May County, an attribute that made the area perfect for aerial maneuvers.

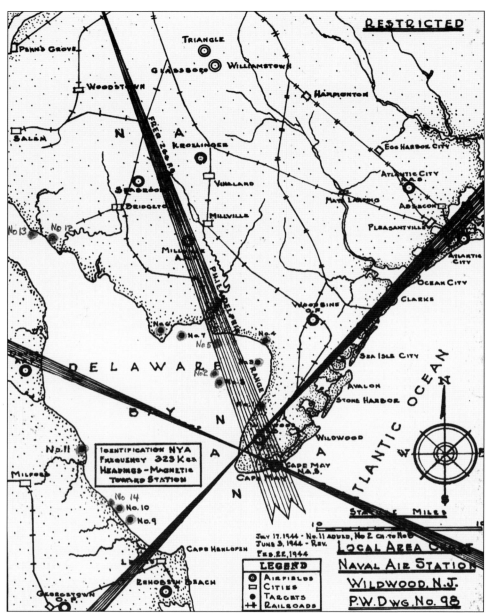

This chart, drawn in 1944, shows 14 separate targets along the Delaware and New Jersey sides of the Delaware Bay that were used during fighter-plane training. Many of the targets were in the vacant, uninhabited land comprised of the marshes or sand dunes that lined the Delaware Bay: two in Cape May County, two in Cumberland County, and one in the state of Delaware. Eight were located in the waters of the Delaware Bay; one of these (no. 5) was deemed not usable for unknown reasons. The majority was used to teach dive and glide tactics, while the remainder was used for antisubmarine, strafing, ship target, and rocket practice. Targets along the county's Atlantic Ocean coast were prohibited because of the public beaches used there during the summer months. That did not stop pilots from flying low over the beaches, however, and tourists often complained of the noise and nuisance they created while doing so.

Line service trucks like this one were a common sight along the aprons at the air station. Equipped with a sturdy lift cranked manually, the truck was used to load bombs onto the fighter planes. Seen behind the truck, with its wings folded, is an SB2C Helldiver; the control tower is visible in the distance. (Courtesy of the National Archives.)

The fixed machine-gun range was sited on the northeast side of the air station, not far from the skeet tower. Made of wood, the open-air range building featured stands on which the machine guns were placed for practice. A modest roof overhead protected the gunners from the sun or rain as they shot towards targets placed in front of a large earthen berm. (Courtesy of the National Archives.)

Two fighter planes, one high in the sky and the other low on the horizon, practice strafing and bombing over a water target set on Kings Pond near the Maurice River in Cumberland County, New Jersey. They are too small to identify but could have been any of several types of fighter aircraft in use at the time by the navy. The location, remote from any houses, farms, and other development, was about 18 miles northwest of the air station, an easy flight over the Delaware Bay. This target was one of several that were added after mid-1944 when the war escalated and the navy needed an increased number of trained pilots to fight in the Pacific theater. A circular target that appears to be fabricated from thick wood stakes is placed not far from the center of the pond. Twin bomb splashes in the water show that the pilot missed the target by several hundred feet. The photograph was taken May 13, 1945. (Courtesy of the National Archives.)

To make the bombing and strafing practice more realistic, the U.S. Navy brought in three out-of-service vessels and moored or sunk them in the Delaware Bay. The largest of these is an all-wood boat that appears to have been a three-masted schooner, stripped of its masts; it probably plied the Atlantic coast around the dawn of the 20th century with shipments of coal, lumber, and other hard goods.

The ships were moored at target No. 13, identified on the target map as being located west of the mouth of Stowe Creek in Cumberland County. This continues to be near an uninhabited area comprised of marshy lowland. The types or kinds of the two small vessels seen here cannot be determined from the photograph.

This close-up photograph, taken May 28, 1945, shows the near-complete annihilation of the three target ships through practice strafing and bombing runs. The fighter pilots who dropped the bombs appear to have been well trained given the amount of damage they inflicted. It is not known if the vessels' bombed-out remains were removed by U.S. Navy personnel or still sit in their watery grave.

Accessed by a dirt road, seen here at the top of the photograph, this target was located near East Creek, the waterway that divided Cape May and Cumberland Counties. Like other target areas, it was sited in a remote location away from farms and in marshlands along the Delaware Bay. Locally abundant clamshells were used to make the circular outlines.

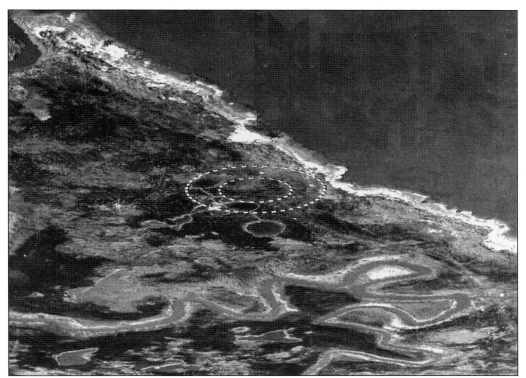

In April 1945, this new target was created and was used for dive-bombing and aerial rocket practice. Its exact location is not known, but it appears to be in coastal lowlands along the Delaware Bay. The rectangles creating the semicircles are six-foot-by-three-foot wood frames covered with white fabric. Fire pots were placed between the frames for night bombing.

The all-purpose range accommodated a variety of firearms. Two rows of metal stands are placed in front of wood or metal frames over which targets were stretched for practice. The area's naturally sandy soil was easily banked into the tall earth berm seen here placed behind the targets to absorb the spent bullets. (Courtesy of the National Archives.)

Unidentified navy airmen take turns as they practice firing a machine gun at the gunnery range in this photograph dated May 30, 1944. The training seat swung around on a limited arc to simulate the shooting conditions in a fighter plane. All of the U.S. Navy's top fighter planes, including Corsairs, Avengers, Dauntlesses, Hellcats, and Wildcats, were equipped with these weapons. They were typically mounted on the aircraft's wings. The Avenger also had a firing turret in the rear of the fuselage manned by a third crewman. A pilot's or crewman's skill in using the machine guns against the enemy might meant the difference between life and death during air-to-air or air-to-ground combat. A stand identical to the one seen in the photograph was donated to the Naval Air Station Wildwood Aviation Museum and is on display there. (Courtesy of the National Archives.)

Driven by a giant propeller powered by what appears to be an airplane engine, seen covered with a shroud in the photograph, this target boat was used to pull an airborne practice target held aloft much like parachutes in parasailing today. The large engine allowed the small craft to go as fast as necessary while towing the target. While land-based targets were easy to simulate, it was much more difficult to replicate the conditions of fighting an enemy warplane while flying. The target boat provided a safe alternative to using one's own fighter planes in practice against each other. An antenna is seen on a stand in front of the motor; it was probably used to communicate with the pilot. The boat was docked alongside the crash boat in the newly built Cape May Canal. Today the former site is near the docks of the popular Cape May-Lewes Ferry in Lower Township. (Courtesy of the National Archives.)

Built almost entirely of dressed stone, the Cross Ledge Lighthouse was first lit in 1875. Located about 25 miles from the entrance of the Delaware Bay, the lighthouse was erected to warn mariners of the location of a submerged ledge that was dangerous to vessels that might stray slightly off course. When two metal lighthouses were built nearby (Elbow of Cross Ledge and Miah Maull Shoal) in the early 1900s, the Cross Ledge lighthouse was no longer needed and it was discontinued around February 1910. It stood abandoned for decades until the military decided to use it for target practice during World War II. Fighter pilots from the air station flew training sorties over the Delaware Bay and dropped small practice bombs on the lighthouse. Amazingly, the lighthouse remained fairly intact after the war but was burned by the U.S. Coast Guard for safety reasons in 1962. (Courtesy of the National Archives.)

This aerial view, taken February 5, 1945, shows several boats moored at the crash boat dock located in what is now the Cape May Canal. The rigorous training over water brought numerous fighter-plane crashes, but they were not always fatal. The crash boats were sent out to pick up survivors and to assist with the recovery of damaged aircraft. (Courtesy of the National Archives.)

A modest snow fell on January 16, 1944, leaving several inches on the runways to be cleared by a snowplow. Despite the inclement weather, training continued and the runways had to be kept clean for the many takeoffs and landings that would be done that day. Snow falling in large amounts was rare in Cape May County, since the water surrounding the peninsula tended to keep the air warm. (Courtesy of the National Archives.)

Four navy airmen practice shooting with rifles at one of the air station's two practice ranges. They are identified, from left to right, as Joseph Rixey, two unidentified, and Jerry Gentili. The practice ranges, one called the fixed machine-gun range and the other the all-purpose range, were located on the northeast side of the air station, well beyond the runways. Skills learned on the range were then honed in a fixed aerial gunnery trainer called the "gunairstructor." Invented by Robert Edison Fulton Jr. during World War II, the gunairstructor was used as a training aid for aerial gunners serving in the U.S. Army Air Forces and Navy. More than 500 were ordered by them during the war, and it is said that Fulton learned to fly in order to write the manual for the trainer. Two buildings on the air station's plot plan are identified as containing gunairstructors, but neither building stands today.

Four

PILOTS, CREWMEN, AND PERSONNEL

The officers and enlisted men of VB-89, a bombing squadron, pose in front of one of their Helldiver fighter planes on December 21, 1944. The squadron was established at Naval Air Station Wildwood on October 2, 1944, and departed for Naval Auxiliary Air Station Oceana on January 1, 1945. They flew SB2Cs on the USS *Ticonderoga* and *Antietam* while fighting in the Pacific theater.

Avis Shervey (right) married Dale Mummert in the chapel at Naval Air Station Wildwood on July 13, 1945. She was a U.S. Navy WAVE, working as a Link trainer at the station; Mummert was in the U.S. Army. The Link trainer was a flight simulator used to train pilots during the war by responding to the pilot's controls and giving accurate readings on the simulator's instruments.

VB-82 is shown in this photograph, taken on April 1, 1944, being commissioned on the apron behind the control tower. VB-82 departed for NAAS Oceana on June 16, 1944, and then left for Naval Air Station Norfolk on September 17, 1944. The squadron was involved in Tokyo and Iwo Jima operations in February 1945 and in the Okinawa campaign on April 6, 1945. (Courtesy of the National Archives.)

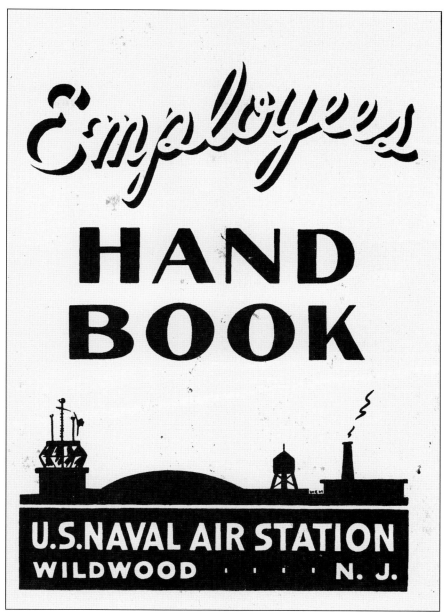

Civilian employees hired to work at the air station received a 20-page handbook that outlined the basics of federal government employment there. All were considered members of the U.S. Navy Civilian Service. Such matters as pay scale, vacation time, sick leave, meals, and promotions were covered in the booklet. By law, they worked 48 hours per week but were paid overtime at the rate of one-and-a-half times their normal hourly rate for time spent after 40 hours each week. Workers were cautioned to maintain secrecy about base operations, including the names, numbers, and functions of the squadrons so as not to jeopardize the war effort. In order to get gasoline coupons from the rationing board, employees had to have at least three riders in their vehicle with them. They were also encouraged to have payroll deductions to purchase war bonds and were free to choose the amount that would be taken out weekly.

An African American navy man loads a bomb onto a dolly while three others inside the munitions magazine assist with the hoisting bar. When the United States entered World War II in December 1941, the navy's African American sailors had been limited to serving as mess attendants, officer's stewards, or ship's cooks for nearly two decades. With the pressures of wartime severely draining manpower resources, however, changes gradually occurred. Although the navy remained racially segregated in training and in most service units, in 1942 the enlisted positions were opened to all qualified personnel. When commissioned in April 1943, Naval Air Station Wildwood's enlisted crew, except for members of the Stewards Branch, were 100 percent white. By May 1945, the station's enlisted crew was comprised of 61 percent white men, 22 percent African American men (one of whom was a gunnery instructor), and 17 percent WAVEs. (Courtesy of the National Archives.)

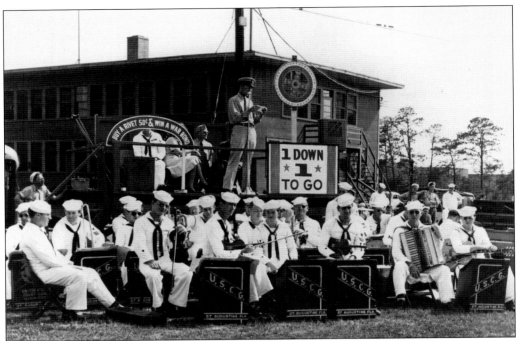

Lt. Cmdr. Morris Brownell Jr., the air station's executive officer from April 1, 1943, to July 1, 1943, solicits donations to the war effort from a podium standing in front of an unidentified air station building. The slogan "1 down 1 to go" referred to the victory in Europe as "1 down" and a potential victory in the Pacific as "1 to go."

VB-14, seen in this photograph, was commissioned June 2, 1943. It merged with VB-15 two weeks later and the squadron's designation was changed to VB-2 on July 16, 1943. The bombing squadron served on the carrier *Hornet* and participated in the attack on Pelieu Islands in March 1944. It also played a major role in the June 20, 1944, attack on Vice Adm. Jisaburo Ozawa's northern carrier group in the Pacific.

On January 6, 1945, the Steward's Mates at the station's Bachelor Officers Quarters posed for this photograph. The overwhelming majority (68 percent) of Steward's Mates were African American, joined by some of Asian descent. They were responsible for serving meals and attending to the officer's needs, service positions that were typical for the time and had been for at least two decades in the navy. Experiments in integration convinced the navy that it was practical, and after the war the navy was the first of the major branches to adopt integration as an official policy. Only towards the close of the war did the navy begin outfitting vessels with predominantly African American crews, and in 1944 the U.S. Navy commissioned its first-ever African American officers. During the war, African Americans also served in construction battalions and serviced airplanes, yet only a small number were given the opportunity for frontline service. (Courtesy of the National Archives.)

Pictured here in September 1944, the officers, pilots, and crewmen of VB-86 pose in front of and on one of the Helldiver fighter planes in their bombing squadron. VB-86 was established at the air station on June 15, 1944, and departed for NAAF Otis on October 2, 1944. The squadron then served on the carrier *Wasp*, fighting in the Okinawa campaign in April 1945; it was in combat on V-J Day, August 15, 1945.

George Samuel (Sam) Pollard (left), a U.S. Navy pilot, is shown here with his crewman in their fighter plane named *Duck Butt*. Pollard met Wildwood-native Florence Changlin while at the station and they married shortly after the war ended. He is at the controls of what appears to be a Curtiss SB2C Helldiver, a dive bomber responsible for more shipping kills than any other aircraft during the war.

The training personnel at Naval Air Station Wildwood pose outside of Hangar No. 2 in this photograph taken April 18, 1945. Lt. Cmdr. R. S. Eberhart, seated in the center of the first row, was in charge. The rest are identified as, from left to right, (first row) Kenneth Bicbal, Dutch Kaufman, Carl Smith, Lt. (jg) Paul Basker, Lieutenant (junior grade) Sorby, Lieutenant Cape, Lt. Keith Manion, Lieutenant (junior grade) Weaver, Ensign Moulsen, Chief Kaufman, Louis Winkler, and Ricardo Palmer; (second row) Jean McKee, Ruth Haskins, Mary Farland, Betty Gough, Ann VanAntwerp, Imogene Williams, Nancy Fielding, Pat Waite, Irene Carlson, Ann Vosberg, Avis Shervey, Penny Bieber, Joan Sydler, and Jo Walhunis; (third row), B. Blackges, W. Stickley, J. Cahill, B. Benham, E. Wilkerson, Grace Christopher, Marilyn Fleck, Betty Westbrook, Bobbie Leighton, Nancy Peterson, Pat Doughery, Roy Stokes, Louis Soul, and William Henry. One of the WAVE trainers, Imogene (Williams) Gluck, was responsible for teaching the operation and use of a machine gun.

On January 15, 1945, Comdr. Arthur Priest Linscott of Brookline, Massachusetts, relieved Comdr. F. B. Connell as the station's commanding officer. This photograph, taken in the station auditorium, shows the changing-of-command ceremony. Linscott received his naval aviator's wings in 1926, served as a flight instructor before World War II, and was executive officer of two other naval air stations before coming to Naval Air Station Wildwood in 1945.

VB-13, seen here in front of Hangar No. 1, was commissioned on November 1, 1943. The squadron was sent to the Pacific on the carrier *Franklin* and was involved in the second Battle of the Philippine Sea in September 1944. One month later, it was the heaviest hit in a battle on October 15, 1994.

William Lindgren, a U.S. Navy pilot from Wernersville, Pennsylvania, poses on top of his Corsair fighter plane. Lindgren served in the navy for four years, flying first Wildcats before moving to Corsairs. While serving on the USS *Sargent Bay*, he was part of fighter group composite squadron VC-79 that was responsible for protecting fuel tankers in the Pacific theater. As the carrier's fighter planes patrolled overhead, the other units of the replenishment groups provided fuel and replacement aircraft to the fast carriers, enabling them to remain at sea for extended periods. In February 1945, the carrier provided air support for landings on Iwo Jima. The squadrons on board also flew artillery-spotting and group-support missions as well as antiaircraft and antisubmarine patrols until the ship departed a month later to provide similar support during the Okinawa campaign. After the war ended, the *Sargent Bay* had the pleasant task of bringing troops home from Hawaii.

Several squadrons gather on April 5, 1945, for the awarding of medals and the inspection of personnel by Commodore Gordon Row. Squadron officers stand in several rows with groups of enlisted navy men behind them. Seen in the background is Hangar No. 2, completed just two months earlier. The one-story buildings adjacent to the hangar were ready rooms used for training-mission briefing and debriefing.

VBF-20 and VB-20 were commissioned on April 16, 1945; this photograph documents that commissioning along with a medal awards ceremony by Commodore Gordon Rowe. Front and center is William Lindgren, a pilot who was also an instructor at Naval Air Station Jacksonville and who taught flight instructors at Naval Air Station New Orleans.

VBF-152 was commissioned on March 5, 1945 inside Hangar No. 1. In the background are two Corsair fighter planes. A bombing-fighting squadron, VBF-152 departed for Naval Auxiliary Air Field Groton on May 2, 1945, and flew many missions on an escort carrier in the North Atlantic. At the time, German submarines were plentiful and the squadron flew most of their missions at night.

Commodore Gordon Rowe, at the microphone, speaks during the commissioning of VB-20 and the awarding of medals on April 16, 1945. A bombing squadron, VB-20 had earlier served in several important Pacific theater battles, but it was reformed in April 1945 at Naval Air Station Wildwood. The squadron departed for Naval Air Station Edenton (North Carolina) on June 22, 1945, and its wartime activities after that are not known.

Standing with Lt. Clarence Doud, head of the communications department, are telephone operators, from left to right, Rita A. Long, Marie Reed, Ida Wise, Olga Larcombe, and unidentified. These female operators were part of a civilian workforce that numbered 386 by April 1945, up significantly from the 10 that were employed when the station was commissioned in 1943. Civilian workers included laborers, firefighters, and maintenance workers.

Designs for targets and a map showing their locations are pinned to the wall behind the desk of J. A. Hain, the navy man in charge of targets and target areas used by the airmen for practice. The U.S. Navy established at least 14 targets over land and water that were sited on both the New Jersey and Delaware sides of the Delaware Bay.

In June 1945, 125 German prisoners of war arrived from Fort DuPont in Delaware and were assigned to Naval Air Station Wildwood. One of their first assignments was to dig ditches around the air station to control the mosquito population. The marshes and slow-moving creeks that characterize much of the county's bay and ocean-side terrain provided perfect breeding grounds for the pesky summer insect. This photograph was taken August 2, 1945, and it must have been a hot and humid day judging by the prisoners' sparse attire. Several of the men are carrying axes and wearing hip-high wading boots as they slash and chop their way to clear the dense thicket growing around a creek. The letters *pw* (standing for prisoner of war) are written on their shorts. Some of them were also assigned to the transportation and public works departments at the station where there was an acute manpower shortage. They were housed in a former World War I army camp close by. (Courtesy of the National Archives.)

Although ditches helped to drain water from low-lying areas, some waterways needed more drastic measures to help rid them of mosquito breeding grounds. Seen here are German prisoners of war building a retaining wall along an unidentified creek in Cape May County, probably one close to the air station. One of the men wears a shirt marked merely with "PW," signifying his status. (Courtesy of the National Archives.)

A navy color guard and a small band of navy men that included one WAVE participate in the commissioning of VBF-152 on March 5, 1945. Hangar No. 1, the size of two football fields, was large enough to accommodate both fighter planes and ceremonies. The hangar's north wall, seen here, looks much the same today as it did then.

On clear, dry days commissioning ceremonies were held outside of Hangar No. 1. This photograph, taken May 1, 1944, depicts the commissioning of VB-83 with Lt. Comdr. Frank A. Patriarca, U.S. Navy, as commanding officer. The squadron was involved in the last major attack on Japanese warships maneuvering in the open sea on April 2, 1945, and was in combat on V-J Day, August 15, 1945. (Courtesy of the National Archives.)

Officers and enlisted men of fighter-bomber squadron VBF-75A pose with one of their Vought F4U Corsairs in front of Hangar No. 2 on June 16, 1945. The squadron reported to the station on June 1, 1945; it was redesignated VBF-75 and departed for Naval Auxiliary Air Station Chincoteague on August 9, 1945. (Courtesy of the National Archives.)

Lt. Cmdr. G. H. Roberts (left) and an unidentified U.S. Navy officer share the podium at the commissioning of VB-97 on November 1, 1944. Roberts was the bombing squadron's commanding officer. VB-97 used SB2C-4Es and at least a half-dozen Corsairs for its aircraft. The extent of their participation in the war is not known. (Courtesy of the National Archives.)

Rear Adm. James O. Richardson (center) inspects the remote-control plane setup at Naval Air Station Wildwood with two unidentified U.S. Navy officers in 1945. Manufactured by the Culver Aircraft Company and designed by Al Mooney, small radio-controlled airplanes known as Culver Cadets or Culver TDC2s were acquired in 1941 by the navy. These all-wood drones were used to train antiaircraft gunners.

Comdr. Francis Ballantyne Connell is shown seated at his desk in this February 26, 1944, photograph. He graduated from the Naval Academy with the class of 1918 and served with the Atlantic Fleet Cruiser and Transport Force throughout World War I. In 1920, he received aviation training at Pensacola and further training by the army as a pursuit pilot at Kelly Field in Texas. He was then assigned to a torpedo-plane squadron at San Diego, and until 1923 he served in all the types of squadrons and on the staff there. Retiring in 1936 from navy service, Connell became vice president of Inter-Island Airways of Hawaii. He returned to active duty shortly before the outbreak of World War II, serving first as air officer for the 12th Naval District and then the 6th Naval District. He then served as Naval Air Station Wildwood's first commanding officer from April 1, 1943, until January 15, 1945, when he was ordered to Port Hueneme in California. (Courtesy of the National Archives.)

Two unidentified navy men walk past one of the station's 136 buildings in this undated photograph. Because most metal was used for war needs, like bombs, aircraft, and ships, the station's buildings were erected mostly of wood and other nonmetal materials. Seen covering the side of the structure behind them are fireproof transite panels made from asbestos and Portland cement.

H. Fisher (left), of the station's photographic office, and chief Roope pose for this undated image. The building behind them cannot be identified, but it is covered with transite panels like most of the station's structures. Transite, advertised as "stone lumber," was introduced as a new building material in the early 20th century and was noted for its durability and resistance to rot and rust.

The officers and enlisted men comprising CASU-23 (Carrier Aircraft Service Unit) pose in front of several Curtis SB2C Helldivers parked on the apron in front of one of the station's hangars. Among other support duties, men in this unit were responsible for servicing and maintaining the aircraft, ensuring that they were flight-ready for the pilots in training.

Navy man Joseph Rixey is shown here standing next to one of the vehicles used at the station to transport the German POWs. He wears a gun in a holster at his waist, but there were no reports in local newspapers of any of the prisoners of war trying to escape.

VF-97
FIGHTER TRAINING SQUADRON

J.C. Sinclair Lt.(jg) USNR.	D. Maculey Lt. USNR.	A.H. Hopkins Lt. U.S.N.R.	C.E. Guilott Ens. USNR.	H. Hills Lt. U.S.N.R.	J.A. Sherrod Lt. U.S.N.R.
C.E. Pringle Lt.(jg) USNR.	C.J. Wendel Ens. USNR.	B.S. Keller Lt. U.S.N.R.	A.R. Coffin Lt.(jg) USNR.	G. Shaw Lt. U.S.N.R.	W.C. Edwards Lt. U.S.N.R.
E.S. Rice Ens. U.S.N.R.	J.J. Joseph Lt. U.S.N.R.	E.J. Callahan Ens. USNR.	R. Peters Lt. U.S.N.R.	A.V. Timer Lt. U.S.N.R.	J. Harris Lt.(jg) USNR.
R.E. Hawkins Lt.(jg) USNR	R.M. Spivey Ens. U.S.N.R.	D.G. Stewart Lt. U.S.N.R.	C.T. Neher Ens. USNR.	A.D. Creek Lt. U.S.N.R.	C.F. Myer Lt. U.S.N.R.
R.B. Cairns Lt.(jg) USNR.	T.L. Falvey Lt. U.S.N.R.	R.L. Martin Lt. U.S.N.R.	M. Lihn Lt. U.S.N.R.	W.E. Eder Lt. Comdr. USN.	R. Hood Lt. U.S.N.R.
R.G. Gray Lt. U.S.N.R.	D.F. Nelson Ens. USNR.	J.R. Perry Ens. U.S.N.R.	R.R. Whitte Lt. U.S.N.R.	B.M. Johnson Lt. U.S.N.R.	C.B. Gray Ens. U.S.N.R.

VF-97, a dive-bomb training squadron, poses at Naval Air Station Wildwood. The group was stationed here for most of the war but spent two weeks at Naval Air Auxiliary Station Manteo, North Carolina, in 1945 when dive-bomb squadrons were transferred there. They quickly returned to Naval Air Station Wildwood because the facilities at Manteo were not suitable.

In addition to digging mosquito-control ditches, the German POWs were also used for a variety of construction projects at the air station. They are shown here helping to lay the foundation for a roadbed in front of the station maintenance shop at the corner of Yorktown and Ranger Roads. They also picked tomatoes for local farmers.

Lt. Albert Timer poses at his desk while stationed at Naval Air Station Wildwood. He spent 38 years in the military, serving first in the National Guard and then joining the navy reserve in 1942. As a navy pilot, he taught instrument flying to both U.S. Navy and Marine Corps pilots. While at Naval Air Station Wildwood, he was a safety inspector and had to insure that each fighter plane was safe to fly before taking off for training missions. He also commanded an antisubmarine squadron at Willow Grove, Pennsylvania. When he retired from the navy, he left having attained the rank of commander. After the service, he worked for Lockheed Aircraft for 25 years, retiring from that company as chief flight test pilot. During his aviation career, he logged over 44,000 flight hours. After the war, he managed a certified FAA inspection station for the Associated Products of America freight airline based at the Cape May County Airport.

A group of unidentified civilians, probably firefighters, and their dog pose with a fire truck in front of one of two firehouses erected at the air station. This structure, the larger of the two, was built with terra-cotta blocks and was sited across from the administration building. It is one of a handful of original buildings still standing.

Navy pilot John Marsh poses in front of his SNJ fighter airplane on the apron at Naval Air Station Wildwood. His flight suit was typical navy-issued gear and consisted of flight pants, a fabric bomber jacket, flight goggles, and a fabric helmet with the headphones built into it. The straps around his chest and legs supported a parachute worn on his back.

Five

CRASH LANDINGS AND OTHER MISHAPS

This Hellcat landed nose first on the runway just north of the control tower and Hangar No. 1. Aside from bent propellers, which were about 13 feet in diameter before the damage, the fighter plane appears to be fairly intact. A group of service personnel watch as one worker prepares to tie a rope to the fuselage from a boom truck in order to right the plane.

The many farm fields that surrounded the naval air station provided a flat but extremely hard surface on which to crash land a fighter plane experiencing difficulties. This Hellcat flipped completely over, damaging not only the propeller, but the wings, cockpit, and tail section as well. It is not known if the pilot survived.

The only sign of damage to this Grumman Hellcat appears to be the loss of its rear landing wheel, which explains why a crane was needed to move it. The engine oil seen dripping over the intact outlet was typical of air-cooled fighter planes. The pistons were arranged in a circle around the propeller's drive shaft, and those pistons on the bottom of the circle always leaked.

It is not known if rainy conditions, suggested by the nearby puddles and overcast skies, played a role in the crash landing of this Grumman F4F Wildcat near the station. Navy personnel have attached a rope across the plane's underbelly and appear to be removing what might be munitions from the tail section.

The wing-mounted landing gear failed to deploy on this Corsair, which led to its crash landing somewhere near the air station. Ever mindful of the possibility of fire, either from the engine, the highly flammable fuel it used, or the armament carried on board, service personnel have positioned a fire extinguisher nearby.

This Corsair took a nose-dive on one of the air station's runways, radically bending the propellers but causing little damage to the rest of the fighter plane. The cause of the accident is not known. The pilot, seen at the far right, watches as the plane is prepared for righting, a wire tether joined to its rear landing wheel. The metal poles on top of the airplane carried an antenna wire that was also strung to a smaller pole mounted on top of the rudder. The fire truck and firefighters wait just off the runway in case the airplane catches on fire. Seen in the distance is a tetrahedron that acted as a windsock, showing the wind direction on the runways. Corsairs were carrier-capable aircraft used primarily during World War II and the Korean War and had the longest production run—from 1942 to 1952—of any piston-engined fighter in U.S. history.

Snow and ice, seen here coating the runway and the ground, may have contributed to the undignified landing of this Wildcat, which appears to have slid sideways off the runway and into a drainage ditch. The man sitting on the side of the cockpit appears to be the pilot, and he seems to be unhurt. A boom crane truck is seen at the left, its grapple hook being positioned to rescue the aircraft. Although roughly as far south as Baltimore, Maryland, Cape May County had its share of significant snows during the winter months and maintained a fleet of snowplows to keep the runways clear so training could continue no matter what the weather. Although training was halted during nor'easters and the occasional hurricane because of high winds, a light snowfall like the one seen in this photograph was unlikely to cancel flight training.

This Wildcat did a belly flop off of one of the runways, apparently when its landing gear failed to deploy. Made by Grumman, the single-man F4F Wildcat was replaced by the faster F6F Hellcat later in World War II but continued to be built to serve on escort carriers where larger and heavier fighters could not be used.

One of the propeller's blades is firmly implanted in the ground where this Grumman Hellcat landed in the middle of a field near one of the station's runways. The Hellcat was fitted with over 200 pounds of cockpit armor to aid pilot survival, had a bullet-resistant windshield, and had armor around the engine oil tank and oil cooler.

Machine-gun ammunition is being stripped from its storage compartment on the wings of this downed Vought F4U Corsair. Most Corsair fighter planes built during World War II carried six .50 M2 Browning machine guns mounted on their wings. The warplane held 2,300 rounds that gave over one full minute of fire from each gun, and each fired in three- to six-second bursts making it an effective fighter in a variety of conditions. Corsairs were also armed with 2,000 pounds in bombs. These armament features, combined with its handling prowess in the air, made the Corsair a devastating weapon against aircraft, ground targets, and ships. Introduced in July 1941, the airplane quickly became the most capable carrier-based fighter/bomber of the war. It had many nicknames, among them Bent-Wing Bird, Bent-Wing Monster, Horseshoe, Super Stuka, Hose Nose, Hog Nose, Whistling Death, and Sweetheart.

Landing nose first onto an air station runway, this Grumman Hellcat appears to have sustained only minor damage except for its mangled propellers. Several service vehicles, including one with a crane, are on the concrete and ready to provide assistance. Chocks keep one of the landing-gear wheels stationary and a fire extinguisher is placed nearby in case a fire erupts.

A navy serviceman or a station fireman hoses down the concrete on the runway where a Wildcat skidded to rest, its landing gear having failed to deploy. The hose snakes back to what appears to be one of the station's fire trucks. The station had two firehouses: one in the same building as the control tower, the other closer to the dormitories and administration building.

The tail of this SBD Dauntless is in shreds, probably clipped off by another airplane or a vehicle as it stood parked with its wings folded. Seen on the underside of the wings are perforated flaps known as air brakes that were developed to eliminate severe tail buffeting on earlier models. The Dauntless was introduced in 1938 by Northrop and was designated SBD when Douglas bought out Northrop shortly afterwards. The fighter airplane was produced with six variants that incorporated self-sealing and larger fuel tanks, armor protection, and a bulletproof windshield. It had a maximum speed of 255 miles per hour with a cruising speed of 185 miles per hour. A total of 5,938 Dauntless aircraft were built between April 1940 and the end of 1944. Its nicknames included Slow But Deadly (a play on the fighter plane's acronym), Banshee, Speedy-D, Speedy-3, Clunk, and Barge.

This Corsair, its gull-like wings and body nearly flattened by the impact of crash landing in a field of brush and scrub pines, is worked on by a crew of U.S. Navy servicemen, seen here removing armament and other items from it. A pool of dark oil appears to be collecting under the fighter plane's powerful engine, an 18-cylinder Pratt and Whitney that was the largest available at the time.

A P47 Thunderbolt, from the Millville U.S. Army airfield in nearby Cumberland County, had a minor collision with an SNJ while on the runway of Naval Air Station Wildwood. The P47 was a bomber escort, used mainly in Europe during the war and was one of the main U.S. Army Air Forces fighters of World War II. Its nickname was the "Jug."

A crane truck prepares to lift a Grumman F6F Hellcat off the side of a runway where it came to rest, presumably after a failed landing. The landing gear appears to have only partially deployed, which might explain the accident. Except for the crumbled landing gear, the rest of the fighter plane seems intact and it is likely that the pilot survived with only minor injuries. The tail end of a flatbed truck is seen to the extreme left; it was probably used to transport the stricken aircraft to the hangar for repair. The Hellcat was the successor to Grumman's Wildcat and major design changes included a low-mounted wing, wider landing gear that retracted into the wings, and a more powerful engine. It had a wingspan of 42 feet, was 33 feet, 7 inches long, and could reach a maximum speed of 380 miles per hour.

A shackle hanging from a boom truck is about to be attached to this Wildcat that appears to have slid off one of the runways. Seen in the far distance is the control tower with Hangar No. 1 behind it. Because Hangar No. 2 is not seen, the photograph was taken before 1944, when that hangar was built.

A Grumman F4F Wildcat has landed in a ditch near one of the runways, its propeller buried in the ground. Immensely popular during World War II, this carrier-based fighter had a 38-foot wingspan and a 14-cylinder radial piston engine capable of cruising at 155 miles per hour with a maximum speed of 318 miles per hour.

Navy service men work on this Corsair, which appears to have landed with one of the forward landing wheels stuck in the up position. The crash landing bent the propellers and probably damaged the deployed forward landing gear that the four men are inspecting. The machine-gun openings—three on the front of each wing—were taped shut or patched over to show if the guns fired or not. A liquid, possibly hydraulic fluid, leaks on to the runway. Two service trucks, one of them with a ladder and the other with a flat bed, wait nearby ready to render assistance. The pilot's folded parachute rests on the damaged wing, a testimony to his survival. The bent-wing design, seen clearly here, was necessitated by the tall landing gear that resulted from the huge propeller needed for the aircraft's 185 miles per hour cruising speed. Production began in 1942 and ceased in 1952.

Because Naval Air Station Wildwood was not served by the railroad, this photograph may have been taken at one of the auxiliary air stations—either Woodbine, New Jersey, or Georgetown, Delaware—used by Wildwood's pilots for training. This Wildcat appears to have crash-landed safely on the railroad tracks, although its landing gear was damaged and its propeller was bent.

Flames leap and shoot high in the air while the nighttime crash of an unidentified fighter plane from Naval Air Station Wildwood burns out of control in this undated photograph. Approximately 129 accidents occurred with fighter planes based at the station during World War II, resulting in at least 42 known deaths.

A Douglas SBD Dauntless lies in ruins amid a forest of trees somewhere in Cape May County. It is not known if the pilot and crewman survived the horrific crash or not. The local newspapers reported the deaths of at least 42 airmen who lost their lives while training at the air station, but the type of aircraft and its identifying numbers were never mentioned in the articles. The Dauntless was designed as a light bomber and reconnaissance aircraft and served with the U.S. Marine Corps, Army, and Navy air-fighter squadrons. It featured a two-man tandem cockpit with dual flight controls and hydraulically actuated perforated split dive brakes. The plane distinguished itself in the Battle of Midway, June 4–7, 1942, when it spearheaded an attack that sank four Japanese aircraft carriers and a heavy cruiser, significantly changing the course of the war. This battle is widely regarded as the most important of the Pacific campaign of World War II.

The crash landing of this SNJ appears to have been one in which the pilot and gunner merely walked away. Seen in the background is Hangar No. 1 with row upon row of fighter planes lined up in front of it. The SNJ and its variants were heavily produced, with a total 15,495 built into the 1950s.

This TBM Avenger came to rest in a patch of marsh grass somewhere along the Cape May County coastline. TBM Avengers were made by General Motors in Trenton, New Jersey; they were torpedo bombers developed initially for the navy and marine corps but were eventually used by several air or naval arms around the world. The fighter plane first saw action during the Battle of Midway in June 1942.

Flames engulf this unidentified fighter plane that crashed during the night near Naval Air Station Wildwood. Nighttime flying was a required component of fighter-bomber training since many war maneuvers would be carried out under the cover of darkness. It is not known if the pilot survived this horrific crash or was able to parachute to safety.

A Howard aircraft (foreground) collided with an SB2C while taxiing down the runway on August 11, 1944. Details of the crash were not made public, but the Howard suffered major damage to its wings, the midsection of the fuselage, and the tail. The Howard seen here was an instrument trainer aircraft. (Courtesy of the National Archives.)

This TBF Avenger crashed on the west side of Naval Air Station Wildwood on August 11, 1944. Although the fighter plane suffered major damage to one wing, it is likely that the pilot and two crewmen survived because there were no reports of their deaths in the local newspapers. The TBF Avenger was one of several fighter aircraft manufactured by Grumman during World War II until December 1943; those designed as TBM were manufactured by General Motors from 1942 until 1945. Throughout 1944, an average of 10 TBMs were built daily, with monthly production reaching a record high of 400 in March 1945. GM converted all of its automobile manufacturing facilities to war materials production, making not only fighter planes, but trucks, tanks, marine diesels, guns, and shells among other items. The conversion of the factories was not easy and new practices that streamlined manufacturing, assembly, and component production were always being developed. (Courtesy of the National Archives.)

An SBD Dauntless that was part of bombing squadron VB-14 caught fire after crashing on the runway on September 15, 1943. It was piloted by Ens. Thomas Anderson Wilde of Venice, California, who lost his life in the accident. Wilde was making a landing when his plane went out of control and dove into the runway. He died three hours later. The small fire extinguishers seen strewn near the wreckage appear to have been no match for the fierce flames fed by the aircraft's fuel tanks. Wilde's squadron, VB-14, arrived at the air station on June 2, 1943, and merged with VB-15 a few weeks later. The squadron's designation was changed to VB-2 on July 16, 1943. Between March and September 1944, the group was assigned to the carrier *Hornet* for combat operations. By this time, the majority of deployed navy SBD squadrons were inshore scouting units whose main purpose was antisubmarine patrol. (Courtesy of the National Archives.)

Because the navy pilots trained over the open waters of the Delaware Bay, a crash boat was kept in readiness should a fighter plane have to make an emergency or forced landing in the water. The crash boat was primarily responsible for rescuing the downed fliers. According to reports, at least 20 aircraft were lost in the Delaware Bay during World War II, with many of the pilots perishing as a result in these training accidents. Some pilots were able to parachute to safety, while the bodies of others were never recovered. The reasons for these losses include mid-air collisions, forced landings in the water, hitting objects (possibly birds) in the air, emergency landings because of engine malfunctions, wings falling off, and spinouts. Missions being carried out at the time of the accident included dive bombing, individual combat, glide bombing, night training, chemical spraying, and rocket training. (Courtesy of the National Archives.)

Six

LEISURE TIME

The air station offered plenty of off-duty activities and put together its own baseball team, which is shown here. The group, comprised of the CASU support members permanently stationed there, played other local men's teams, but it is not known if they had a winning or losing season the year this photograph was taken.

U. S. NAVAL AIR STATION
WILDWOOD, NEW JERSEY

The Officers' Club cordially invites you to attend a Supper Dance in celebration of the First Anniversary of the commissioning of U. S. Naval Air Station, Wildwood, New Jersey.

Buffet Supper, 2000 until 2130.   1 April 1944
Dancing, 2100 until 0100.
$1.50 the person

The Officers' Club was the site for many off-hours functions, including a special "Supper Dance" held to celebrate the air station's first anniversary on April 1, 1944. At $1.50 per person, this affair was more expensive than most, but it offered a buffet dinner that was served from 8:00 p.m. until 9:30 p.m. and dancing from 9:00 p.m. until 1:00 a.m.

OFFICERS' CLUB
U. S. NAVAL AIR STATION
Wildwood, N. J.

Commissioning Anniversary Supper Dance

1 April 1944
Buffet Supper, 2000 until 2130
Dancing, 2100 until 0100
Ticket of Admission

No. 41

These two tickets commemorate the air station's first anniversary supper dance that was held in the Officers' Club on April 1, 1944. The Officers' Club originally stood in the center of the air station, just north of the mess hall where it was easily accessed from most of the station's streets.

Navy airman Paul L. Hebert (seated right) enjoys a drink and appetizers with fellow airmen and others at the American Legion building in Wildwood, New Jersey. The American Legion building was several miles from the air station in the seaside resort of Wildwood, but it was within walking distance of the Hotel Davis, which was used to house navy men during World War II.

The Officers' Club was christened on March 2, 1944, and this photograph commemorates that day of its opening. The club was located on, but not accessible from, Breakwater Road, which formed the southernmost boundary of the air station. Navy officers are seen here in the foreground raising their glasses in a toast to the new facility.

Many of the air station's social events were held at the American Legion building in Wildwood, a resort town located on a barrier island about 5 miles to the southeast. Erected after World War I, the American Legion building featured the meeting hall seen here, decorated with emblems of various fighting units. Note the 48-star flag in the case on the wall.

Nothing is known about Bill Hammersley, who formed a band of U.S. Navy men that performed during socials and dances at the American Legion building in Wildwood. He may be the saxophonist, who was an officer, seated behind the stand with the name on it. Behind him, hanging on the wall, is a photograph of Gen. John J. "Black Jack" Pershing, the renowned army general who served in World War I.

The Officers' Club at Naval Air Station Wildwood served as the setting for this formal dance. The officers' dark suits suggest it was a fall or winter event, but the exact date is not known. The room, with its wood-coffered ceiling and curtained windows, was more elegant than most found in the buildings at the station.

A large group of military and civilian employees from Naval Air Station Wildwood gather for a social in the meeting room of the American Legion building in Wildwood. Navy WAVEs at the air station, many of them shown in this photograph, played an important role in day-to-day operations there, helping to train the airmen and performing many of the clerical duties in the office.

The beach in Wildwood provided the setting for Memorial Day festivities on May 30, 1945. This tribute to U.S. Army and Navy heroes was fashioned from a lifeguard boat decorated with flowers and greenery and topped with an American flag. A large crowd gathered on the sand as well as on the boardwalk seen in the background of the photograph. Although the air station was located some 5 miles to the west of its namesake, a contingent of its officers were billeted in a hotel there and many of the airmen frequented the resort's restaurants and bars during their off-duty hours. Navy pilots also flew over the resort's beaches—often too low according to the newspapers—to impress their wives or girlfriends who were there during the summer months. At the time, Wildwood was a summer resort best known for its amusement piers and boardwalk. Today it is still an active resort. (Courtesy of the National Archives.)

Seven

AFTER THE WAR

Lt. Cmdr. Livingston Ackert oversaw the decommissioning of Naval Air Station Wildwood after the war ended in 1946. The photograph is inscribed to Margaret Harris Reget, who worked in the administrative clerical section of the air station from 1943 until it closed in 1947. She then transferred to Naval Air Station Atlantic City and worked there until 1958.

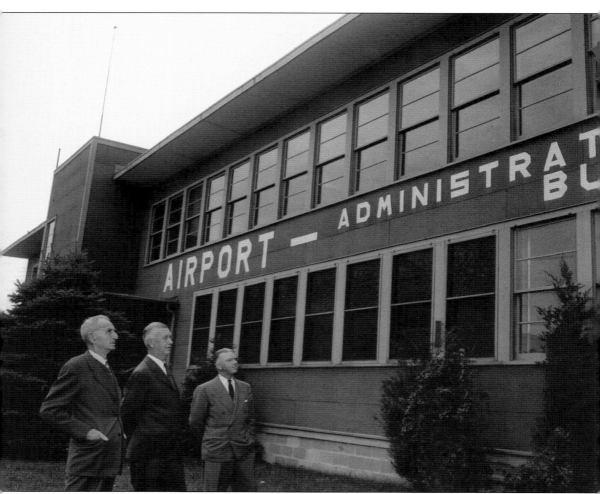

Cape May County freeholder Walter Treen (right) of Wildwood and two unidentified men stand in front of the airport administration building shortly after the station was decommissioned and ownership reverted back to the county in December 1947. The air station's administration building was then used to house the airport's administrative functions. It originally stood at the corner of Lexington and Ranger Roads and was demolished in the 20th century. Treen was placed in charge of the county airport and was modestly successful in creating an industrial park there in those former Navy buildings that were not sold as surplus and moved off the site. One of the buildings that remained was the sprawling mess hall, the third largest of the original structures. Still standing but altered from its original appearance, it has had many industrial uses over the years and in the 1960s and 1970s was used for etching glass bottles. The building is vacant today.

Looking over the facilities at the county airport during the 1950s are, from left to right, William Piper, president of Piper Aircraft Corporation; Walter Treen, county freeholder; and E. Cooper, with Bell Telephone. Piper, often called the "Henry Ford of Aviation," is best known for manufacturing the J-3 Piper Cub, a small single-engine airplane that was used to teach more Americans to fly than any other aircraft at the time. During World War II, four out of every five U.S. military pilots got their primary training in a J-3. The inexpensive Cub had a reputation for being easy to handle and forgiving of a pilot's mistakes. At the time this photograph was taken, Piper was looking to relocate his manufacturing efforts into a larger facility to increase production. However, he chose property at the Vero Beach (Florida) Municipal Airport over property at the Cape May Airport in 1955.

Part of the Navy Day celebration held October 27, 1945, included an air parade featuring various airplanes. Three SNJs, manufactured by North American, are seen here flying low over one of the station's four runways. They were single-engine aircraft used to train fighter pilots serving in the U.S. Navy and Army Air Forces.

Navy Day is an annual event held in October since 1922 to recognize and commemorate the U.S. Navy and those who have served in it. Naval Air Station Wildwood was still operating in October 1945, the date of this photograph, and hundreds of visitors have gathered to watch an air parade that featured many of the war's most successful fighter planes.

Navy men and visitors alike gather on Navy Day, October 27, 1945, to watch TBM-3E Avengers taxi down the runway. These torpedo bombers became one of the three most numerous carrier aircraft of all time by war's end and participated in every major air-sea battle of the war between June 1942 and August 1945.

The apron on the north side of Hangar No. 1 was crowded with airplanes and spectators on Navy Day, October 27, 1945. The front row of parked aircraft featured more than 20 Corsairs, recognized as the best fighter plane used by the navy in World War II. They continue to be featured in air shows today and are easily recognized by the unique, inverted shape of their wings.

Three SNJs fly maneuvers over the east-west runway during Navy Day held at the air station on October 27, 1945. The training plane, which could reach a maximum speed of 205 miles per hour, was known as the "pilot maker" because of its important role in preparing pilots for combat. Over 17,000 were built, and it is estimated that about 350 are still airworthy today.

A Curtiss SB2C Helldiver (center) prepares to take off during one of several aeronautic exhibitions at Navy Day on October 27, 1945. Guided by signals given by a signalman standing on the runway, the pilot would get to the proper position and then drop the plane's wings in preparation for takeoff. The Helldiver could carry 1,000 pounds of bombs or an internal torpedo.

A TBM Avenger, parked on the apron in front of Hangar No. 2, attracts a crowd of onlookers on Navy Day in October 1945. The Avenger's torpedo delivery capability had a huge impact on the Japanese fleet during World War II, and it entered service in the U.S. Navy just in time to participate in the Battle of Midway in June 1942.

Three Corsairs fly low over one of the air station's runways during Navy Day on October 27, 1945. Led by Lieutenant Dibbles, they were part of several flight exhibitions that day, which included dive-bomb attacks, formation flying, and a smoke-screen demonstration. The Corsair's huge propeller allowed the plane to reach an air speed of 400 miles per hour.

Two Helldivers prepare for takeoff in the center of the runway as visitors watch from the sidelines during Navy Day held at the air station on the afternoon of Saturday, October 27, 1945. Although the war had ended, several squadrons and a full complement of fighter planes were still based at the station. Over 2,000 people attended the event, which featured an air demonstration begun by Avengers that were joined by Hellcats and Corsairs; they rendezvoused in the air, circled, and then flew over the air station in attack formation. The air parade was followed by aerial maneuvers performed by seven SB2C Helldivers, led by Lt. Cmdr. Anthony Schneider, holder of two Navy Crosses and numerous other decorations for successful attacks in the Pacific against enemy shipping and shore installations. Also participating were SNJs, led by R. M. Witmer, a pilot credited with shooting down five Japanese planes and holder of two Distinguished Flying Crosses received for fighting against the Japanese at Okinawa and Tokyo.

The Christmas dinner served in 1945 was probably a festive one with the war over and most bomber and fighter squadrons decommissioned. For those still stationed at Naval Air Station Wildwood, the celebratory meal promised to be a hearty one. The menu cover, printed in red, green, and black, features sailors with white collars and hats. The inside of the menu, printed with black type on a red background, shows they feasted on the traditional roasted turkey prepared with oyster stuffing; the oysters were probably harvested locally from the waters of the Delaware Bay. Roast pork was also served, with mincemeat pie, fruitcake, hard candy, ice cream, and nuts offered for dessert.

1945
Christmas Dinner

U. S. NAVAL AIR STATION
WILDWOOD, NEW JERSEY

Commanding Officer

A. P. LINSCOTT

Commander, USNR

Executive Officer

A. R. TAYLOR

Commander, USNR

Supply Officer

M. H. GATCHELL

Lt. Commander, USNR

Chief Commissary Steward

COOKS — BAKERS — BUTCHERS

MENU

Tomato Juice Cocktail

Turkey Noodle Soup

Roast Young Tom Turkey

Chilled Cranberry Sauce   Oyster Dressing   Giblet Gravy

Roast Loin Pork, Apple Sauce

Mashed Irish Potatoes   Candied Sweet Potatoes
Buttered Fresh Asparagus   Baked Creamed Style Corn

Lettuce and Tomato Salad, Mayonnaise

Mince Meat Pie, Whipped Cream   Fruit Cake
Ice Cream   Hard Candy   Assorted Nuts

Parkerhouse Rolls

Bread   Butter   Coffee

Cigars   Cigarettes

In 1954, the county airport retained most of the air station buildings that were not sold at auction. The largest of these structures included Hangars No. 1 and No. 2, the mess hall, the administration building, the aviation utility shop, one of the storehouses, and the operations building with control tower. Although the layout of the station's streets remained, the only visible reminder of the buildings that once lined them are the faint lines in the photograph marking where their foundations once stood. Today only a handful of original buildings remain, and many of them were altered from their original appearances as they were converted for industrial uses during the mid- to late 20th century.

The federal government sold off 79 of the station's 126 buildings as surplus in the summer of 1947. Structures disposed of included the dog kennel, storehouses, ammunition lockers, huts, gate towers, the beauty parlor, the morgue, and one wing of the mess hall. The buildings had to be moved to a location off the station and many, most highly altered from their original appearance, are scattered today throughout the county.

After World War II, the air station reverted back to county ownership and became the Cape May County Airport. A man, identified as only Jerry, stands in front of his Aeronca airplane on the airport apron in September 1953. Aeronca Aircraft Corporation produced training, observation, and liaison aircraft during and after the war. This one appears to be the highly popular Champion model.

United States Overseas Airlines occupied Hangar No. 1 from 1949 to 1964. A privately owned airline, it was started by Dr. Ralph Cox, a dentist who served as a U.S. Navy flyer during World War II. The airline successfully serviced many destinations including Hawaii, New York, Chicago, California, Paris, and China with a perfect safety record.

Ralph Cox purchased war surplus airplanes when he started his own airline, rebuilding and repairing them when necessary. When finished, they were painted and given the airline's logo in a stripe down the fuselage. By 1962, the airline maintained six DC6s and 12 DC4s that were capable of flying anywhere in the world. This undated photograph shows one of the airline's airplanes being painted.

In 1951, United States Overseas Airlines flew hundreds of drums of Aldrin, an insecticide, aboard three of its cargo airplanes to southern Iran. The country was plagued with one of the worst locust invasions in 80 years, and the United States agreed to provide the country with crop-dusting airplanes that were disassembled for the flight and the insecticide, which had been used successfully in the states to combat grasshopper invasions.

A four-engine DC4, made by the Douglas Aircraft Company, has its cargo unloaded in front of Hangar No. 1 in this photograph that was probably taken in the 1950s or 1960s. Versions of the DC4, designated R5D, were used by the U.S. Navy during World War II. After the war, DC4s were used as long-range passenger airliners and were favored by charter airlines.

Rescued from a dump site, a former Naval Air Transport Service airplane sits in Hangar No. 1 while being rebuilt by United States Overseas Airlines (above). The transport service was created shortly after the attack on Pearl Harbor when the timely delivery of equipment and personnel to various far-flung naval commands was vital to the war effort. By war's end, the fleet had grown to more than 431 aircraft within 13 squadrons. No longer needed after the war, many were sold to small, private airlines and converted for passenger service like this one. The totally remodeled aircraft, seen on the runway (below) outside of Hangar No. 1, looks nothing like the transport plane it was originally.

U.S. Overseas Airlines, Inc. Christmas Party, 1952

The location of the 1952 Christmas party held for the employees of the United States Overseas Airlines and their spouses is not known, but it appears to have been well attended. Around this time, the airline was one of Cape May County's largest employers with many people working not only at the company's headquarters at the county airport, but some stationed in remote parts of the world as well.

Forty-three employees of United States Overseas Airlines pose in front of this four-engine DC4 in 1952. Parked inside Hangar No. 1, it appears that the airplane was being refurbished at the time. The size of two football fields, the hangar could easily accommodate the repair and maintenance of several large airplanes at one time.

United States Overseas Airlines had an extensive machine shop set up in Hangar No. 1. The company also manufactured several pieces of equipment in the early 1950s that were used to test the engines in a twin-engine jet bomber being built for the federal government. The new bomber was the Canberra 57A model, developed during the Korean War, and now known as the B-57.

An unidentified worker for the United States Overseas Airlines prepares to cut a piece of wood in one of the workshops located in Hangar No. 1. Because the airline manufactured aircraft component parts and often rebuilt salvaged airplanes, even those that had crash-landed and been cannibalized for their parts, a complete woodworking and machine shop was absolutely necessary to the company's success.

The interior of Hangar No. 1 became the main base of repair and maintenance operations for United States Overseas Airlines. It is seen here on the rare occasion when no aircraft were being stored or worked on inside. The size of two football fields, the hangar was ideally suited for its postwar use for a privately owned, nonscheduled airline company.

United States Overseas Airlines altered the two-story office section that fronted Hangar No. 1 by adding a third story, seen here covered with dark-colored shingles. A projecting bay window added at the same time overlooked the control tower and gave a panoramic view of the airport. The smoke is coming from an unidentified fire not too far away.

By 1996, historic Hangar No. 1 had been abandoned for several years and was left to decay. Windows were broken or missing altogether, the exterior siding was deteriorating, and it had a hole in the roof approximately 50 feet by 100 feet. Dr. Joseph Salvatore and his wife, Patricia Anne, formed a nonprofit organization that purchased the building, listed it in the National Register of Historic Places, and began restoring it.

Hangar No. 1 has been restored to its *c.* 1944 appearance and now serves as the home for the Naval Air Station Wildwood Aviation Museum, an appropriate use for the historic, giant wood structure. Grants and a variety of other funding sources helped to finance repairs to the roof, the installation of new windows, and restoration of the telescoping doors among other things.

INDEX

About Naval Air Station Wildwood Foundation and Aviation Museum

Naval Air Station Wildwood Foundation is a nonprofit organization whose mission is to restore Hangar No. 1 at the Cape May Airport, Cape May County, New Jersey, and to create an aviation museum honoring the 42 U.S. Navy airmen who died while training at the station during World War II.

In June 1997, Naval Air Station Wildwood Foundation purchased Hangar No. 1 at the Cape May Airport. The 92,000-square-foot, all-wooden structure was in a state of disrepair and required extensive renovation. Under the stewardship of Naval Air Station Wildwood Foundation, the hangar was listed in the New Jersey and National Registers of Historic Places and is considered nationally significant for the role it played during World War II.

Naval Air Station Wildwood Aviation Museum now boasts over 26 aircraft displays as well as exhibits of military memorabilia, engines, photographs, and more. Additionally, the Franklin Institute of Philadelphia donated a wealth of interactive exhibits that allow visitors to discover the science of flight. The museum features a library, food vending area, and a recently expanded gift shop.

In its role as a community resource, Naval Air Station Wildwood Aviation Museum hosts activities including fly-ins, aviation festivals, big band concerts, swing dances, veterans' ceremonies, historical lectures, school field trips, and senior tours.

The museum is open daily during the spring, summer, and fall; call for winter hours.

You are invited to become a member of Naval Air Station Wildwood Aviation Museum. Your membership supports the mission of the foundation, which is a tax-exempt 501 (c)(3) organization. Membership entitles you to the following: free admission to Naval Air Station Wildwood Aviation Museum, 10-percent discount in the gift shop, advance invitations to special events and museum functions, reduced admission to special events, and issues of aviation museum quarterly newsletter *The Osprey*.

Please help us give the past a future and become a Naval Air Station Wildwood member! Visit our website http://usnasw.org.

Naval Air Station Wildwood Aviation Museum
500 Forrestal Road
Cape May Airport, NJ 08242
(609) 886-8787 (609) 886-1942 (fax)

www.arcadiapublishing.com

Discover books about the town where you grew up, the cities where your friends and families live, the town where your parents met, or even that retirement spot you've been dreaming about. Our Web site provides history lovers with exclusive deals, advanced notification about new titles, e-mail alerts of author events, and much more.

Find Your Place in History.